# Wax Extraction for the Back Yard Beekeeper

Dave Atherton

NB

Northern Bee Books

First published 2021
Published in the United Kingdom by
Northern Bee Books,
Scout Bottom Farm,
Mytholmroyd,
West Yorkshire HX7 5JS
Tel: 01422 882751
Fax: 01422 886157
www.northernbeebooks.co.uk

ISBN 978-1-912271-94-8

Design and artwork, DM Design and Print

*My heartfelt thanks go to all those beekeepers,*
*who taught me all I know, and who sorted me out*
*when things went wrong.*

*There are too many to name here,*
*but you know who you are.*

## Wax Extraction

Some beekeepers don't recover wax. They regard it as a lot of work for insufficient return, and just throw it away.

At the other extreme, some nerds glean every bit of brace comb and add it to the wax harvest. I'm a nerd.

I started beekeeping in 1994, and soon encountered the problem of recovering wax from old frames. Beginner's mistakes create this situation early on in his or her beekeeping career.

First I tried using an old Burco boiler. For today's youth, this was a glorified electric kettle in which you boiled the soiled terry-toweling baby's diapers, in the days before disposable nappies. You cut the old comb from the frames and put it into an old pair of tights to contain the debris, weight it down under the water, and switch on. Theoretically, the wax melts and rises to the surface and solidifies when cool. In practice, the results were not great. Hours of boiling produced only small quantities of dirty wax, with enormous consumption of electricity.

Next I tried building my own steam extractor using a wallpaper steamer and an old brood box, together with a mesh grid and collecting tray. This was even less successful, although I suspect I didn't sufficiently persevere with the project.

I was horrified by the sinful waste of electricity during these two attempts. After all, you are only trying to melt wax. It takes a lot of energy to heat water as well as the wax, and even more to convert water at 100°C to steam at the same temperature. It is no accident that the Industrial Revolution was powered by steam.

Sunshine is free, so in the late 1990s, I resolved to build a solar extractor. However, I live in the north west of Ireland, and collecting enough heat from the sun can be a challenge. Therefore the extractor had to be well insulated, with double-glazing, and capable of being easily moved around and oriented, in order to catch the best of the sun at all times of the day. When I am melting wax, I frequently adjust the angles of the extractor to get the best result.

Before writing this description, I checked a few beekeeping books, but most of them barely mentioned wax recovery, or gave very sketchy accounts. Dave Cushman's website, normally a good source of information, is disappointing. I turned to Google and YouTube, and there are dozens of clips on the subject. But again, most of them still fall very short in one way or another.

Most of this Internet stuff is from places with unlimited sunshine – America, Australia, New Zealand and southern England. Frankly, some of the equipment described in these clips is laughably inadequate unless you live in the tropics.

Here in the north west of Ireland, May is the best month of the year for sunshine, and the best days are those with a clear blue sky after a rainy spell to clear the air. The warmer days of June to August are also useful, but can be cloudier or hazy, with less collectable radiation. I have photovoltaic panels on my house roof feeding electricity into the mains grid, so I have good data on the amount of sunlight.

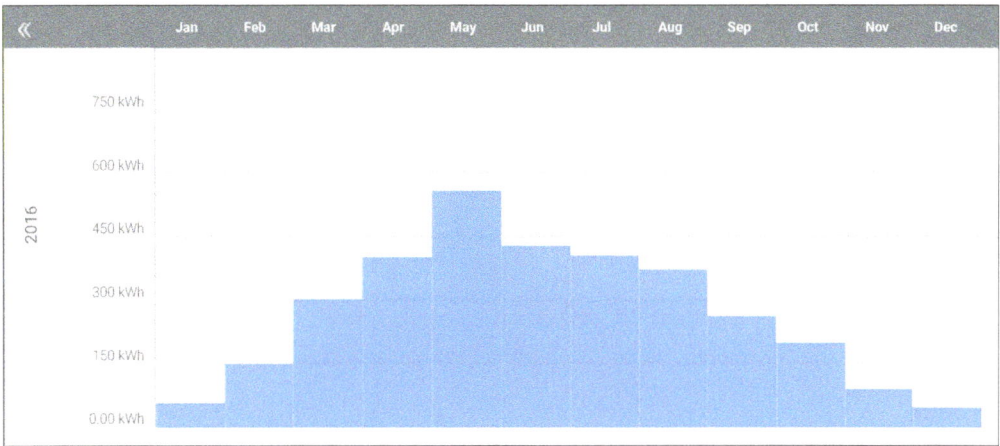

*Figure 1*. **Sunshine**

Kilowatt-hours of electricity, generated per month by my PV roof panels over one year. The pattern varies from year to year, especially June/July/August, but May is consistently the best.

There are two components to my extractor – the solar panel itself, and the carriage on which it is mounted. The carriage is equally important as the solar panel, because without easy manoeuverability, the solar panel can't achieve its full capability. Most descriptions of solar wax extractors completely miss this point.

Building the whole machine did indeed take a lot of time, trial and ingenuity. It worked well for over 20 years, but by 2020 the carriage had become tired, so I used the spring Covid lockdown to build an entirely new one to a different design. The solar panel itself received only minor changes and a lick of paint.

Was it all worth the effort? Possibly not. Some people just use an old oven, and since wax melts at such a low temperature, not much electricity is consumed.

## Solar Panel

My panel is the size it is because I had some scrap materials lying around. I'm a bit of a squirrel. Most of the components of my extractor were left over from other projects, or scrounged from somewhere. I had an old wooden-framed single-glazed window and a sheet of metal from some appliance or other such as a fridge or washing machine. I can't remember what that appliance was, but the sheet had shallow turned-up edges; and it just fitted behind the window frame, and was wide enough to hold a National or Commercial brood frame. Using planks from a fence I demolished, I built a wooden box to hold the sheet of metal, with 60mm polystyrene foam insulation at the back, sandwiched by plywood.

To double-glaze it, I routed a recess in the window frame and bought a sheet of 6mm glass to fit, long enough to run rainwater off the panel. This glass was the only significant expenditure. It was bedded in silicone mastic.

Double-glazing units are readily available nowadays. Recently I spotted a heap of rejected ones outside an industrial unit. "The hungry eye sees far."

# Carriage

Getting the geometry of the balance point and position of the elevating mechanism was difficult, with many false starts and re-starts. I just experimented and fiddled until it was satisfactory. The problems arose because, as the elevation increases, the centre of gravity changes relative to the balance point.

The pivot was made from a length of threaded bar, sleeved with plastic central heating pipe and located by recycled door bolt retainers. The elevating mechanism was based on the scissors jack from a Fiat Strada. (Sure, everybody has this stuff lying about the place, don't they?) The Strada went to the scrap yard a decade before I even thought about beekeeping, but you never know when you are going to need a scissors jack.

The following illustrations show it far better than I can describe in words.

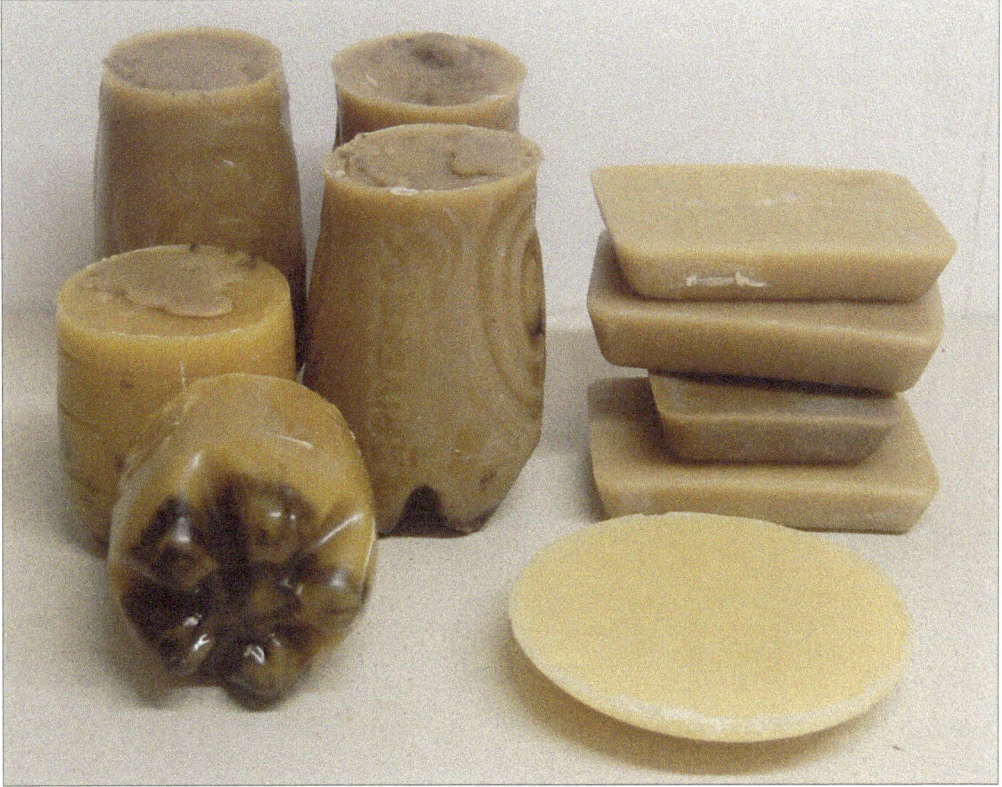

*Figure 2.*

Rendered wax ready for exchange for foundation.
The light wax at right was recovered from cappings, using gentle heat to avoid discoloration [see page 25 (v)]. It is being prepared for show.

*Figure 3.* **General view**

(i)    The rectangular base is made of 4" x 2" (100mm x 50mm) pressure treated timber, with half-lap joints at the corners, glued and through-bolted to the castors.

(ii)    The A-frames are also treated timber, again half-lap jointed at the apex, which is reinforced and weather-proofed with a strip of aluminium attached with stainless steel screws.

(iii)    The lower end of the outer glass overhangs in order to throw off rainwater. 6mm was far too heavy and expensive. 3mm would be sufficient, although I once was glad of the extra thickness when I had an almost catastrophic event. (Long story!)

(iv)    Note the bumper to protect the edge of the glass in the event of a traffic shunt at rush hour. It is made of 1" aluminium water bar.

*Figure 4.* **General view (open)**

(i)  Note the overhanging roof, or ridge tile, which keeps out the rain when closed. This is made from a sheet of 5mm plastic, bent using a hot-air gun. All materials and design were chosen such the extractor could be left outside all summer. And it rains all summer, right?

(ii)  The frames hang on long spouting bolts mounted on the metal backing sheet.

(iii)  The sides of the box were ¾" (19mm) treated timber. Originally they were not insulated, but in the 2020 refurbishment I added an additional plank, which can be seen here.

(iv)  Were I to make a new extractor, I would make the box deeper, to accommodate more layers of frames. Mine only holds two layers, ie. 6 brood frames, or rather more super frames. It would be useful to be able to process a whole brood body of frames in one session.

*Figure 5.* **Minimum elevation.** (The 1-metre rule gives a general idea of size.)

(i)  At minimum elevation, the extractor can catch the last rays of the sun. You would be surprised at how much extra wax can be melted off the frames by a last gleam at the end of an otherwise dull day.

(ii)  The two rear wheels are swiveling castors, and the front ones fixed. I chose fixed wheels because of the experience of wayward supermarket trolleys, but next time I would have all four swiveling, for better maneouverability.

(iii)  My yard is concrete, but if your ground is soft, you would need larger, wider wheels to avoid bogging down.

(iv)  Note the thermometer. 75°C melts much of the wax, but 90°C to 100°C gets noticeably more, and leaves the frames cleaner for recycling.

(v)  The short copper pipe on the side of the panel is a gun sight to help setting bearing and elevation. No shadow equals maximum heat collection.

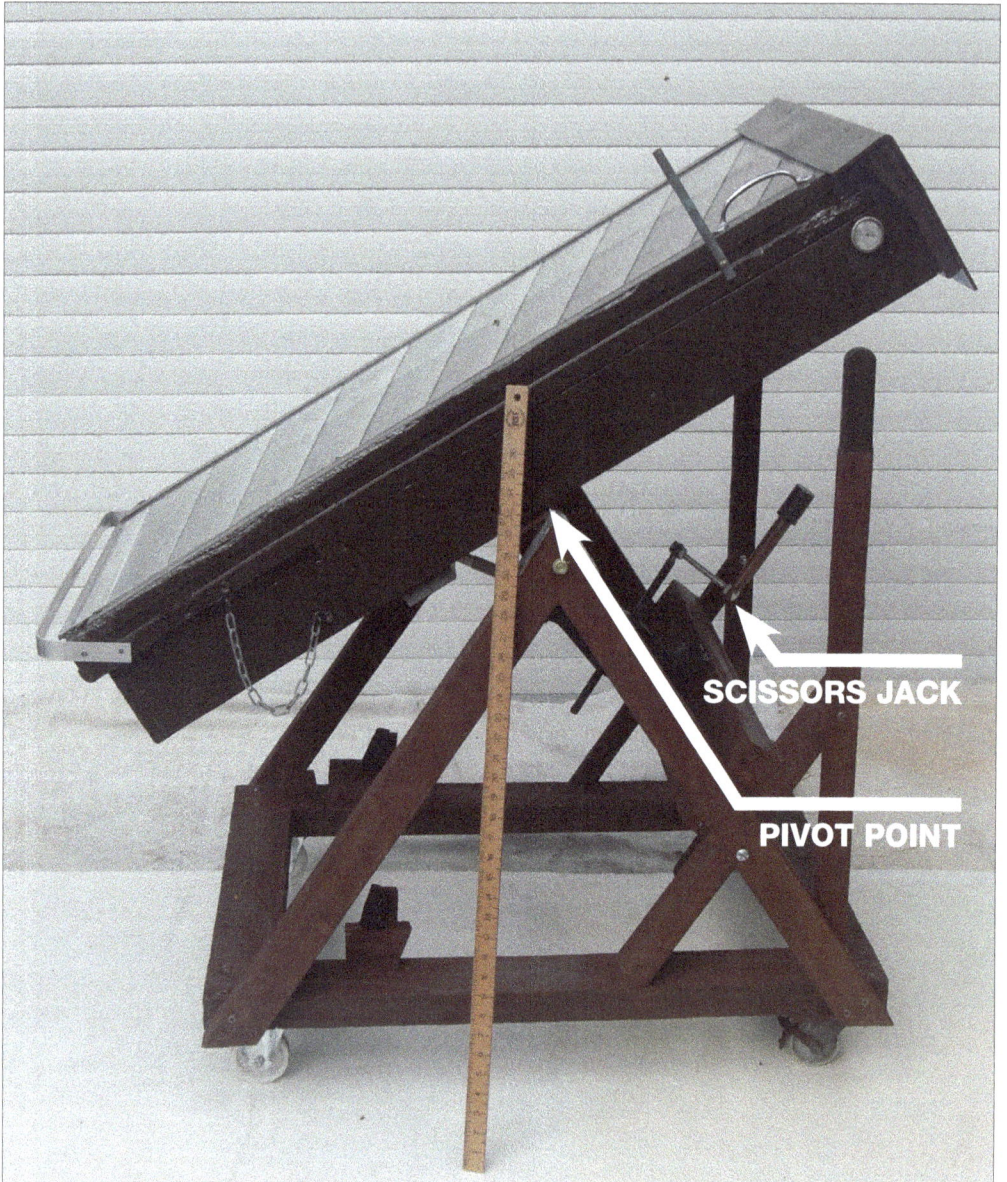

SCISSORS JACK

PIVOT POINT

*Figure 6.* **Maximum elevation.**

Maximum elevation must be sufficient to ensure that the panel is at right angles to the sun at mid-day, at mid-summer.

There is the problem of matching the required lifting force to the load, at different elevations.
As the solar panel is jacked up from the minimum to the maximum elevation, weight is gradually transferred from below the pivot to above. This means that the force required from the jack becomes less as the elevation increases.
Unfortunately, this is the exact opposite of what a scissors jack delivers – perfect for a car, which needs more force as it is raised off its springs, but less than ideal for the purpose in hand.
Fortunately, the scissors jack I used had sufficient power to cover the range, but it took a lot of experimentation to achieve the right balance. Nevertheless, the end result was very satisfactory.

There are two possible answers to the balance problem, should anyone wanting to build a similar extractor, using a weaker jack or other lifting device:

(i)     Suspend the solar panel at its sides, not underneath, so that it is more evenly balanced

(ii)    Alternatively, transmit the power from jack to panel via a pivoting lever, thereby reversing the direction of travel. I leave it to my ingenious readers to make this a reality.

*Figure 7.* **Elevation mechanism.**

(i)     The bar connecting the jack to the panel, and associated brackets, are in aluminium with stainless steel bolts and screws.

(ii)    The crank is made from a coach bolt and off-cuts of copper central heating pipe. (15mm inside 22mm, for strength.)

(iii)   The handle is a 35mm film canister, with the head of a 10mm coach bolt set in Isopon inside. The coach bolt is liberally greased and inserted into the copper pipe. Perfect, a joy to use!

*Figure 8*. **Wax collection.**

(i)     The lower flange of the metal backing sheet was worked into a spout so that the molten wax drips into a 2-litre plastic bottle.

(ii)    A bent strip of metal catches most of the debris.

(iii)   The bottle is held in place by an arrangement of cup-hooks and string. Simple, and very effective.

(iv)    The selection of wooden blocks is to place under the bottle according to its exact height. These bottles shrink slightly when hot, allowing the wax to trickle everywhere!

*Figure 9*. **Solid wax.**

(i)   When cutting off the top of the bottle, leave a tongue as shown, to tuck behind the spout of the metal backing sheet.

(ii)  After the wax has cooled and solidified, the plastic bottle can be cut off with a craft knife. Make a vertical cut down to almost the bottom, and then cut half-way round each side. (Be careful!).

(iii) The remnants of plastic can still go into the recycling bin.

(iv)  There may be some debris on the bottom of the block of wax. Remove it with a hot-air gun and scraper.

(v)   Beeswax becomes discoloured when heated above 85°C, so mine is often rather dark. Wax from old combs is even darker, of course, but mine has never been rejected when submitted in exchange for foundation.

*Figure 10.* **Other features.**

(i)     Note the storage facilities for stout gloves (upper right) and a scraper (lower left). Freshly rendered frames are very hot, and if the day is good, you need to change them quickly for a new batch. Also, give the hot frames a quick preliminary scrape to remove most of the debris, cocoons and wire (if you are going to recycle them).

(ii)    The two upright steering handles are necessary because the whole unit is quite heavy.

*Figure 11.* **Recycling Old Frames**

You've melted the wax off the frames – what are you going to do now?

After the wax has been recovered, the frames can be re-used by scraping and scorching them. Mind you, I once timed myself doing this work, and found that the savings I made for the hours spent, meant that I was paying myself slightly less than the statutory minimum wage. So when I tire of this laborious work, I use the rest for firelighters.

I scorch the frames to prevent disease transmission, but of course, if the frames are the result of a diseased hive, I burn or bury them without even recovering the wax.

Some may argue that frames should not be recycled anyway, but all I can say is that I have been recycling frames for years, and have never had a serious outbreak. I've never seen foulbrood, for example.

*Figure 12.* **Scraping frames.**

These are the tools I use to recycle frames from which I have melted the wax.

(i)    The work surface is an offcut of kitchen worktop. Note the strip of metal screwed to the upper left hand side, necessary because the edge was crumbling after years of knocking off bottom bars.

(ii)    In the lower row of tools, note the rampin.
Useful, but only if the wood is soft.

(iii)    The cranked scraper is excellent for cleaning out the sidebar groove, and the small screwdriver gets the last bit of debris out of the corner.

(iv)    The pincers deal with recalcitrant gimp pins.

(v)    When scorching a hive body, I use the long home-made scraper on the right to deal with frame runners.

*Figure 13.* **Knocking off the bottom bars.**

(i)     Prize off the wedge from the top bar with the chisel.

(ii)    Knock off the bottom bars as shown. One sharp blow should do it. Leave the sidebars in place.

(iii)   Scrape clean every surface of all four components. This is when you are glad that you reached a high temperature in the solar extractor to capture all the wax, and gave the hot frames a quick scrape into the wheelie bin, to remove the bulk of the cocoons etc.

(iv)   When scraping the bottom bars and wedge, first leave the gimp pins in place. When you have done whatever is convenient, knock the pins back into the wood and then scrape the rest of the piece. In this way, it's easy to re-assemble the frame with new foundation using the same pins without removing them.

*Figure 14.* **Ready for scorching.**

(i)     Scorch every surface of every component. Hold the component at one end
        and scorch all sides of one half. By the time you have scorched all
        four components, the first one will be cool enough to hold, in order to
        scorch the other half.

(ii)    Assemble the frame with new wax.

You will soon get into an efficient rhythm, although as I said,
the financial returns are small.

www.ingramcontent.com/pod-product-compliance
Lightning Source LLC
Chambersburg PA
CBHW040154200326
41521CB00019B/2608